S0-CYG-518

BIRD MIGRATION

All animals migrate. A migration is any planned journey from one place to another. This book describes some interesting and extraordinary migrations, including one which starts and finishes at opposite ends of the Earth.

MIGRATIONS

BIRD MIGRATION

Liz Oram

and

R. Robin Baker

Department of Environmental Biology
University of Manchester

STECK-VAUGHN
LIBRARY
A Division of Steck-Vaughn Company
Austin, Texas

Library of Congress Cataloging-in-Publication Data

Oram, Liz, 1964–
Bird migration / Liz Oram and R. Robin Baker.
p. cm. — (Migrations)
Includes index.

Summary: Discusses the migration patterns
of such birds as sparrows, pigeons, owls, chickadees,
swallows, albatross, and penguins.

ISBN 0-8114-2925-3

1. Birds—Migration—juvenile literature. [1. Birds—Migration. 2. Birds—Habits and behavior.] I. Baker, Robin, 1944– . II. Title. III. Series: Oram, Liz, 1964– Migrations.

QL698.9.069 1991 91-12120
598.252′5—dc20 CIP AC

Cover: *A flock of snow geese about to take flight.*

Typeset by Multifacit Graphics, Keyport, NJ
Printed in Hong Kong
Bound in the United States by Lake Book, Melrose Park, IL

1 2 3 4 5 6 7 8 9 0 HK 96 95 94 93 92

Contents

Introduction

Everyone has watched a flock of birds flying overhead and wondered where it was going, or where it had come from. The answer depends on the season of the year, and the type of bird.

Many birds make amazingly long journeys each year. The greatest

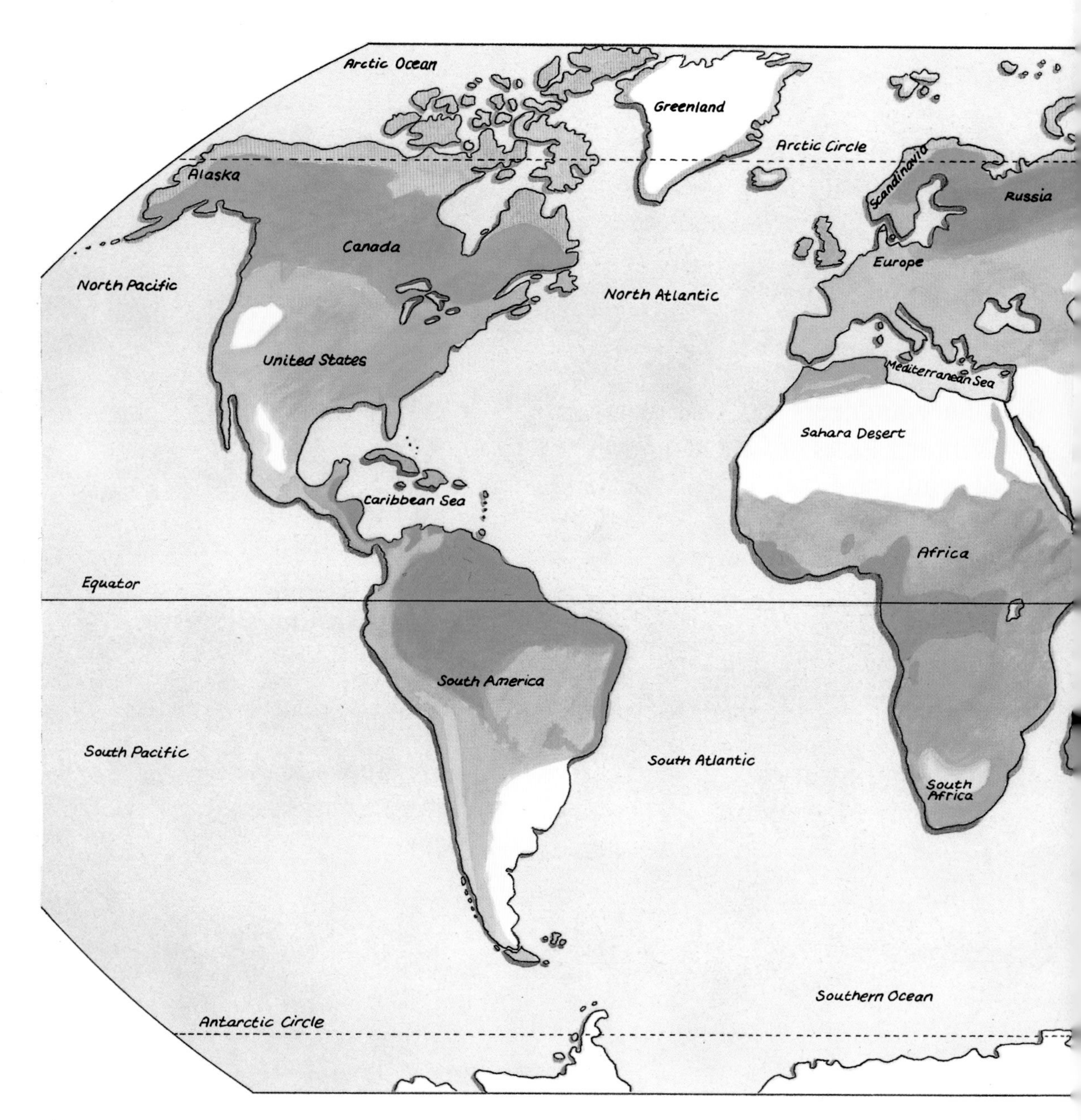

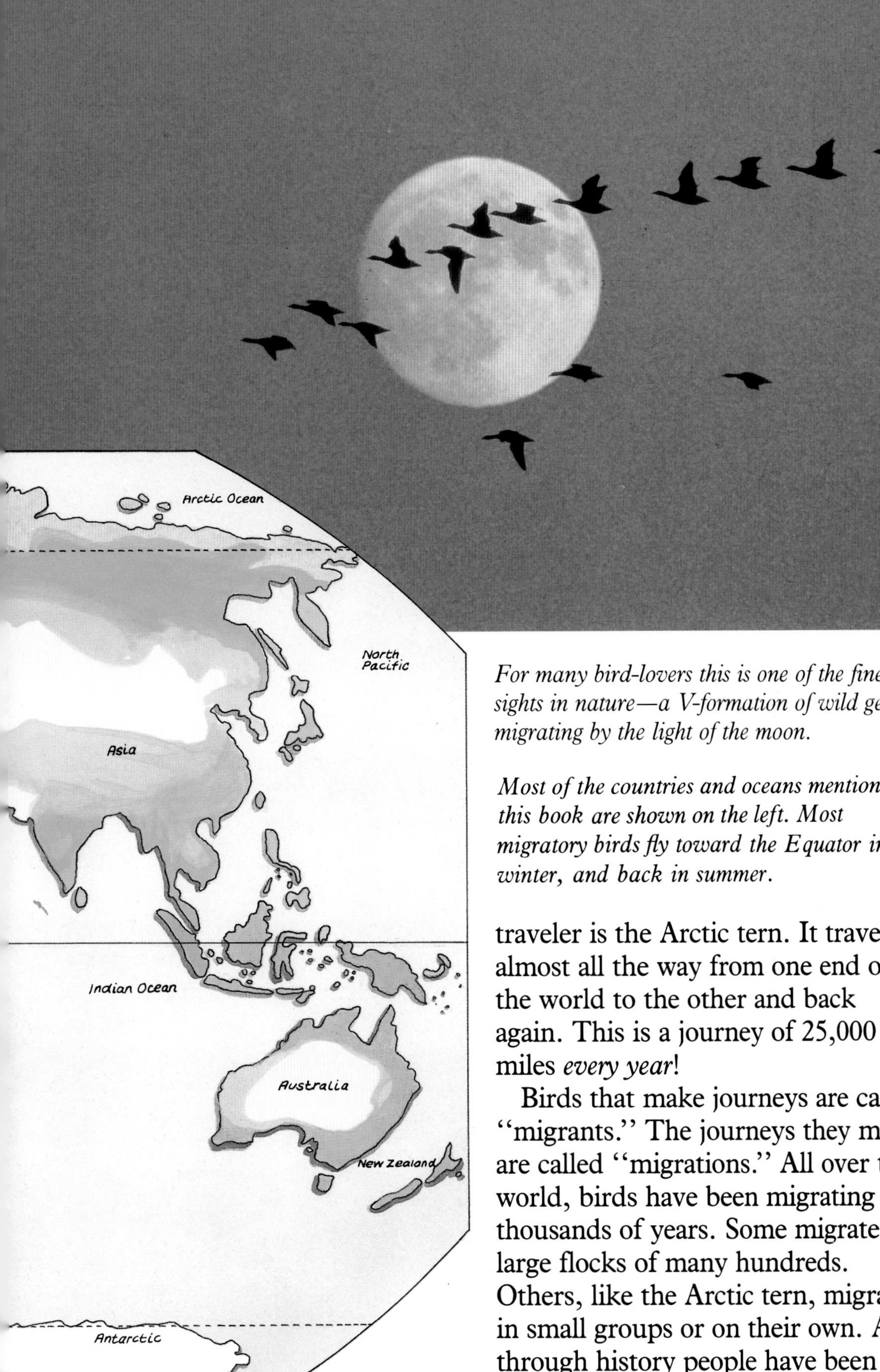

For many bird-lovers this is one of the finest sights in nature—a V-formation of wild geese migrating by the light of the moon.

Most of the countries and oceans mentioned in this book are shown on the left. Most migratory birds fly toward the Equator in winter, and back in summer.

traveler is the Arctic tern. It travels almost all the way from one end of the world to the other and back again. This is a journey of 25,000 miles *every year*!

Birds that make journeys are called "migrants." The journeys they make are called "migrations." All over the world, birds have been migrating for thousands of years. Some migrate in large flocks of many hundreds. Others, like the Arctic tern, migrate in small groups or on their own. All through history people have been

Geese migrate by night as well as by day. Birds take turns leading this family group of pinkfoot geese in formation. No one knows why geese honk as they migrate.

interested in and observing migrating birds.

Not all birds migrate as far as the Arctic tern. Some birds migrate only very short distances. For example, have you noticed all the different birds that visit your yard or park? Most of them are simply looking for food or something to drink, and may never travel farther than a few miles in their whole lives. Yet even these very short migrations are of some interest.

Birds have a wonderful ability to find their way. Some birds migrate thousands of miles, then return to the same tree they started from. In the last chapter we explain how they manage to do it.

1 Birds with a Fixed Home

Most birds are able to migrate long distances. However, many birds hardly travel any distance at all during their lives. In this chapter we look at two birds that hardly travel at all—the house sparrow and the common pigeon.

The house sparrow and the pigeon both have homes where they stay year round. They make tiny migrations in order to feed, but rarely go farther than a few miles from their homes in their entire lives. Most of you have seen these two birds. They are found in almost all parts of the world, both in cities and the country.

The House Sparrow

Most of this bird's feathers are either a gray or a brown color, and this gives it a rather dull appearance. However, the sparrow makes up for this with its warm, friendly personality. More than any other bird, the sparrow enjoys being near people. Almost everywhere that there are people, there are sparrows, too. Sparrows often build their nests in the roofs of houses or other buildings. In winter many people leave bread and other food out in their yards for sparrows to eat.

The house sparrow is unique among birds; it lives only near humans. In winter it is largely dependent on household scraps.

Throughout the ages our affection for this gentle little bird has often been shown. For example, when people first traveled from Europe to colonize North America, Australia, and New Zealand, they took the house sparrow with them. On arrival, the sparrows were released into the wild. The sparrows that now live in North America, Australia, and New Zealand are descended from the sparrows of Europe.

Some sparrows, then, migrated from Europe to places as far away as New Zealand. Unlike most birds, though, they did not migrate by flying. They were carried in ships.

Pigeons are also happy to live near people. In cities, humans supply all their food. They like to build their nests on the ledges of buildings.

Left to themselves, sparrows would not travel very far.

Perhaps there are sparrows living near your house. If so, they will be there all year round. Sometimes they will visit your neighbors' yards. Sometimes they might even fly to nearby parks or fields. Most of the time, though, they will stay within a mile or two of your house and come back every night to sleep.

Pigeons

Last time you visited the center of a large city did you notice all the bluish-gray pigeons flying around? Perhaps you had to sidestep them a few times when they got under your feet. Or perhaps you remember being startled or having to duck when one swooped near your head.

Like the sparrow, the pigeon also makes its home among buildings. Pigeons often build their nests very high up on the window ledges of tall buildings. The nests look so precariously balanced that it is a wonder they do not fall off onto the street below.

Where did these pigeons come from that now live in cities? A thousand years ago their ancestors, known as rock doves, used to live near cliffs by the sea. Like today's pigeons they built their nests very high up, on cliff ledges. (A few wild rock doves are still found in Scotland

and in the region of North Africa.)

In time, people discovered that rock doves were good to eat. They began to keep them in pens, the same way they kept chickens. Every so often a pigeon would be killed for the table. Today the pens are called dovecotes, or pigeon lofts. The pigeons are free to fly out of the lofts, but always return to the lofts at night to sleep. A few pigeons, though, fail to return. The pigeons in cities are the offspring of these escaped pigeons.

The racing pigeon can be taken in a closed box hundreds of miles from its home and will almost always find its way back soon.

Homing Pigeons

Soon after people began to keep pigeons in lofts, an exciting discovery was made. Pigeon owners found that if they took one of the birds a few miles away and let it go, it could fly in a straight line back to the loft. Often the pigeon would arrive home before its owner! This ability of pigeons is now called "homing." Scientists are still trying to find out exactly how the homing pigeon "homes." Pigeons can sometimes home even if they are taken 600 miles away.

People have used the homing ability of pigeons to send messages back to their homes. When an owner goes away, he takes a few of his birds with him. When he wants to send a message back home, he writes a letter and attaches it to a bird's leg. This pigeon then flies back to its loft at the owner's home. The people there catch the pigeon and read the message. Pigeon post is sometimes much faster than mail.

Pigeons are not the only birds that have been used by people in this way. In ancient Rome the swallow was often used as a messenger bird. It carried messages from Rome to people in the surrounding countryside. The frigate bird, which nests on the islands in the Pacific Ocean, is also a messenger. Even today it is still being used by people to send messages between the different islands.

2 Moving to a New Home

Birds like to have a home. Birds with a home know exactly where to go to find food or drink. Sometimes, though, nearby food becomes scarce. What does the bird do then? It has to make its home in a new place where there is food.

In the last chapter we looked at pigeons and house sparrows. These birds are rarely short of food. They live in yards or cities where people leave food around. Pigeons and sparrows hardly ever need to relocate to find more food. Some birds, however, like the crossbill or the jay, often have to move to find more food. These are just two of the birds we will look at in this chapter.

Most migratory birds go back to their original home in summer, but the crossbill—if forced to seek food in another country—rarely returns.

The Crossbill

Crossbills are found all across the northern parts of North America, Europe, and Asia. They live in large fir tree forests. The crossbill gets its name from its unusually shaped beak. The upper and lower jaws are twisted so that they actually cross over at the tip. You may think that such a bill would make feeding difficult. It doesn't. In fact, it is a great help to the bird. This is because crossbills feed on the seeds which are found in pinecones. The crossed bill is perfectly designed to help the bird extract the seeds from inside the cone.

Each crossbill makes its home in about a two square-mile section of pine forest. In most years, the trees in this area will grow enough cones to keep the crossbill well fed with seeds all year long. However, sometimes the weather is so bad that the trees produce hardly any cones at all. In that case, if the crossbill stayed put, it would almost certainly die of starvation. It has no choice but to leave its patch of forest and search for food.

Often, hundreds of crossbills from the same region are forced to migrate together. They may have to migrate long distances before they find a pine forest with plenty of cones. Some crossbills migrate all the way from Scandinavia and the Soviet Union to fir tree forests in France, Britain, and Germany. These new forest homes have to provide enough food to feed the crossbills throughout the whole winter.

Once they have migrated, these birds hardly ever return to the forests of their birth. Usually they live in their new homes for the rest of their lives.

Waxwings and Jays

These two birds also live in forests, and feed on seeds and berries. Jays particularly like oak trees and acorns. Waxwings and jays are found in

The migrations of waxwings are unusual. They normally wander in small groups but sometimes migrate in a mass. They even shift their breeding grounds from year to year.

many parts of North America, Europe, and Asia. When there are not enough seeds or berries for them to eat, they have to migrate to new homes in other forests. Every few years parts of the United States are invaded by flocks of blue jays looking for a new home. Countries such as Britain are invaded by large numbers of pretty waxwings for the same reason.

The Snowy Owl

The snowy owl is a wonderful and mysterious bird. It is large, and is covered with beautiful white feathers. It lives in a part of the world known as the Arctic Circle. The Arctic Circle is a big area around the North Pole. It includes parts of Canada, Alaska, the Soviet Union, and Scandinavia. The Arctic Circle is an extremely cold place. There is snow on the ground throughout most of the year. The summers are short and not warm at all. In fact, the weather in the Arctic Circle is so cold, snowy, and windy that trees cannot grow there. The plants that do grow are short and lie close to the ground. This type of vegetation is called "tundra."

The Ghostly Hunter

The favorite food of the snowy owl is a type of rat that lives on the tundra. This rat is known as the lemming. (Lemmings have very interesting migrations of their own. You can read about them in *Mammal Migration.*)

The snowy owl regularly hunts lemmings. It sometimes hunts at

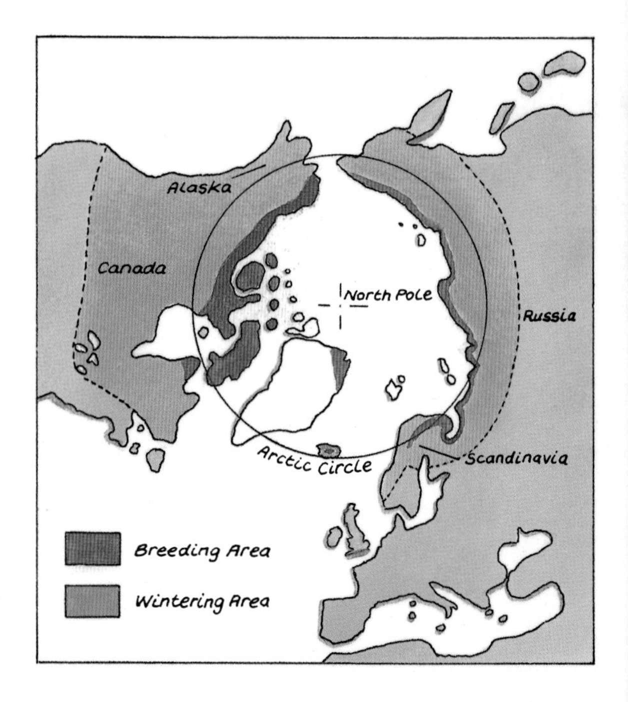

This map shows the usual extent of the snowy owl's range. However, at times when food is scarce, they have been seen as far south as California, Bermuda, and France.

night. In the moonlight it flies silently over the tundra, casting a ghost-like shadow on the ground. Like most other types of owl, it has a marvelous sense of hearing. It uses its sharp ears to detect the tiniest movements of lemmings under the snow. Having located its prey, the owl swoops down silently and picks up the animal with its powerful talons.

In some years there are lots of lemmings living all over the tundra.

Although snowy owls eat a variety of small animals and birds, their favorite food is the lemming. When lemmings are scarce, snowy owls migrate farther south than usual.

Each snowy owl chooses an area where there are enough lemmings for it to eat. Each bird lives in its own area of tundra and raises a family.

However, eventually the supply of lemmings starts to decrease. The small plants on which the lemmings feed grow slowly in the cold weather. After they have eaten all the existing plants, there is nothing else to eat, and many of them starve. This means that there are not enough lemmings around for the snowy owls to eat. Like the crossbills, the owls have to migrate to new homes where there are fresh supplies of food.

Some snowy owls migrate to California and the Carolinas from the Canadian tundra and live on ordinary rats and mice. Owls from Norway have been known to migrate to Scotland. One family of snowy owls that migrated from Scandinavia to the Shetland Islands stayed in its new home for over ten years.

3 The Wanderers

Do you, or does someone you know, own a parakeet? If so, you probably think of them as just pretty little birds that spend all their time in cages. However, not all parakeets live in cages. Wild parakeets live in large flocks, and migrate huge distances in search of food. The parakeet is one example of a group of birds known as "wanderers."

A flock of parakeets often runs out of food every two or three days.

An albatross chick covered with downy feathers. Once it begins migrating the albatross will never stop except to breed.

When this happens the birds have no choice but to migrate to a new area. These birds spend a large part of their lives "wandering" over large distances, from one feeding ground to the next.

Parakeets may fly thousands of miles each year. Albatrosses travel even farther, as we shall see. Other wanderers do not have to travel so far. We'll start by looking at a bird that in North America is known as a chickadee. In Europe it is known as a titmouse. These birds wander around only within a forest.

Winter Flocks of Chickadees

In America there are two types of chickadees—the Carolina chickadee and the black-capped chickadee.

In Europe and Asia there are several types of titmice: blue titmice, great titmice, crested titmice, long-tailed titmice, and several more. They feed mainly off the eggs and caterpillars of butterflies and moths, but will eat many other insects.

In spring and summer there is no shortage of young insects for the chickadees to eat. So, at that time of year the birds do not need to go far for their food.

In winter, life becomes much more

difficult for the chickadees. Their food supply decreases as insects become more difficult to find.

If a chickadee remained in just one area during this time it would almost certainly starve. So it increases its chances of finding enough food by wandering.

However, chickadees do not wander alone. They form large flocks to help each other find food. In Europe or Asia, one flock might contain several types of titmice.

Sharing the Food

When the flock settles to feed, each type looks in a different place for food. If the flock landed in a tree, the long-tailed titmice would search the tips of the twigs for insect eggs. Blue titmice would concentrate on the branches. Great titmice would search the ground underneath.

In winter, flocks of chickadees are a familiar sight in many forests and yards. Next time you go for a winter walk in the woods, look up at the trees. Most of the time you will hear only the faint rustling of wind blowing through the trees. There may not even be a bird in sight. Then suddenly the trees will be full of noise and movement. Look up and you will see lots of tiny birds hopping from branch to branch and frantically chirping at each other. This is a flock of chickadees. After a few moments, the flock will move on and the woods will be silent again.

Many people like to hang up baskets of nuts to attract the pretty blue titmouse to their yards. However, the main food of titmice and chickadees is insects.

Budgerigars

Wild budgerigars, a type of parakeet, live in Australia. They move around in huge flocks that may contain thousands of birds.

The central regions of Australia are very dry. It hardly ever rains; in some places it rains less than once a year. However, when it does rain, the plants grow quickly. After the plants have flowered they produce seeds.

Seeds are the mainstay of the budgerigars' diet. Flocks of budgerigars wander around looking

Budgerigars do not look as if they are migrating, but in their search for water they might travel several thousand miles each year.

for places where it has rained recently. In such places there is no shortage of seeds to eat. Often there is such a good supply of food that the budgerigars can stay for a few weeks. This is long enough for them to build nests, lay eggs, and raise offspring. Sooner or later, though, the food always runs out and the flock is forced to move on in search of another food source.

In their endless search for rain and food, budgerigars may travel thousands of miles each year. Some people think it is cruel to keep a bird that is normally so active and energetic in a tiny cage all its life.

The Wandering Albatross

The wandering albatross is an immense bird. It measures about six feet from one wingtip to the other. Six feet is about the height of a tall man. The albatross spends its life gliding over the waves of the sea looking for fish and squid. It snatches its food at the surface of the water, or just below it.

The albatross lives in the region of Antarctica. Because of the extreme cold, not many people live there. At the center of Antarctica is the South Pole, and all around is ocean. Here is where the Pacific, Atlantic, and Indian oceans come together. This "Southern Ocean" is the hunting ground of the albatross, reaching as far north as the southern tips of Africa, South America, Australia, and New Zealand.

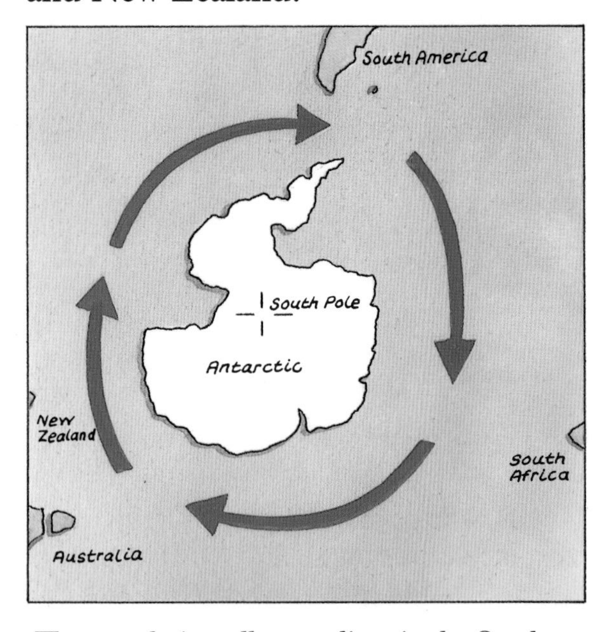

The wandering albatross lives in the Southern Ocean. It migrates ever eastward, passing its birthplace many times during its life.

The wandering albatross has a wingspan of up to thirteen feet. It covers vast distances by downward glides and soaring upward.

Dotted around the Southern Ocean are many cold, windy, and mountainous islands. Every two years many albatross chicks are born and raised on these islands.

As soon as they can fly, they set off on their first migration. The first migration of a young albatross may last as long as five years. Amazingly, the bird hardly ever settles during this five-year period. Its powerful body and immense wingspan allow it to glide tirelessly over the waves in search of food.

Albatrosses have often been known to follow fishing boats for days at a time. To many fishermen the albatross is known as the "sailors companion." It can cover amazing distances. Recently, an albatross on Crozet Island, southeast of Madagascar, was fitted with a tiny radio transmitter. Its flight was tracked from a satellite. In a single month it covered 10,000 miles.

In the cold and desolate Southern Ocean the strong wind always blows in an easterly direction. The albatross almost always flies with this easterly wind. It migrates eastward, eastward, and still eastward around the Southern Ocean, until it finally ends up back where it started. So, having migrated thousands of miles in a big circle, the young albatross returns to the island where it was born. It settles there for a while and raises a chick of its own. Within a year, though, it is wandering the ocean again.

4 Moving with the Seasons

For a bird, life in the summertime is usually quite pleasant. The weather is warm and there is plenty of food. Some birds eat only leaves and fruit. Other birds eat seeds. Some birds eat rats and mice or other small mammals. Whatever a bird eats, there is usually a lot of food around in summer, and the long days allow plenty of time to look for it.

Winters, on the other hand, are cold and cloudy. Moreover, the days are short and the nights are long. The plants stop growing, and many lose their leaves. All the insects and small mammals hide away from the cold. It becomes difficult for birds to find enough to eat. Moreover, the days are so short that most birds have little time to look for food.

In this chapter we will look at a few birds that have two homes. They have one home for the summer, and

Migration is not always north-to-south. In summer these snow partridges live 16,000 feet up in the Himalaya mountains. In winter they migrate downhill to the warmer valleys.

another for the winter. At the end of summer they migrate to the winter home where there is plenty for them to eat.

Partridges in the Himalayas

The Himalayas, between India and Tibet, are the highest mountain range in the world. Many kinds of birds live there. One of them is the partridge.

If you've ever climbed a tall mountain, you've probably noticed that it is much colder and windier at the top than it is at the bottom. The tops of the Himalayas are extremely cold; even in summer the snow doesn't melt. However, halfway down the mountains the weather is much less cold, and in summer the area is alive with plant and insect life. Here is where Himalayan partridges make their summer homes. There is enough food to feed both the parents and their family of chicks.

In winter, though, as it turns colder and colder, snow settles farther and farther down the mountainside. It becomes very difficult for the partridges to find food, so they migrate to their winter homes. They move down the mountainside to the warm valleys at the bottom. In these valleys the partridges find enough food to last them through the winter. In spring, as the mountain snows melt, they start to migrate back up the mountain to their summer homes. Each partridge's two homes may be several miles apart.

The Robin

This bold little bird with its bright red breast is easily recognizable by both American and European children. The robin is found in most

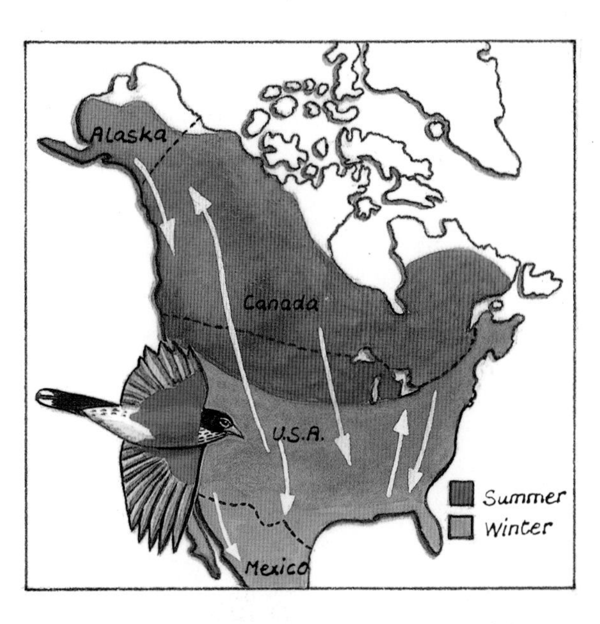

This map shows the migrating routes of the American robin. In general it flies south to warmer weather on the Gulf of Mexico.

parts of North America and Europe, but the British robin is slightly different from the others. British robins have only one home where they stay year round. This is because British winters are not usually very cold or snowy. Robins can find enough worms and other food to survive through the winter.

In parts of North America, northern Europe, and Scandinavia,

In Britain it is rarely very cold, so robins stay through the winter. The American bird also does not migrate from the southern U.S.

where the winters are cold and snowy, robins cannot find enough food to eat. They solve that problem by migrating. Every autumn, these robins migrate to places with warmer winter climates.

Some robins simply look for a protected swamp or cedar grove near their summer homes. Robins from the northern United States and Canada generally fly south to the Gulf coast and Mexico. Most German and Scandinavian robins migrate southwest to Spain and Portugal, although some go to Britain. In these places there is enough food for the birds to stay well-fed and healthy until the following spring. When spring comes, they migrate back to their first homes. These robins have both summer homes and winter homes. The two homes can be as much as 500 miles apart.

Geese and Swans

In Chapter 2 we looked at the snowy owl, a magnificent bird that lives on the tundra near the Arctic Circle. Snowy owls are not the only birds that live on the tundra. Each year, during the summer months, many geese and swans make their homes there. In summer the tundra is almost a paradise for such birds. As the ice and snow melt, the ground becomes dotted with ponds and lakes. In these ponds and lakes small plants and animals grow rapidly. The swans make their homes on the water, and feed on this plant and animal life. Geese eat the small green plants that spring up on land at this time.

The geese and swans build their nests and lay their eggs on the cold ground. Goose and swan chicks are able to feed themselves as soon as they hatch. In this way they are different from the chicks of most other birds. They also have big appetites, and in the following months they grow rapidly. By the end of the summer most of them are able to fly.

A Family Migration

As winter approaches the snows gradually return to the tundra. The ground is covered with a thick coating of snow, and the ponds freeze over. Soon there is no food for the birds to eat. The cold weather kills the green plants, and the swans cannot reach the food beneath the frozen ice. Both geese and swans would die of starvation if they stayed. So, like the robins that spend summers in the United States and

This map shows the main migratory routes of geese and swans. From their breeding grounds in Arctic Canada, Greenland, and Asia, they fly south in winter to the U. S. and Europe.

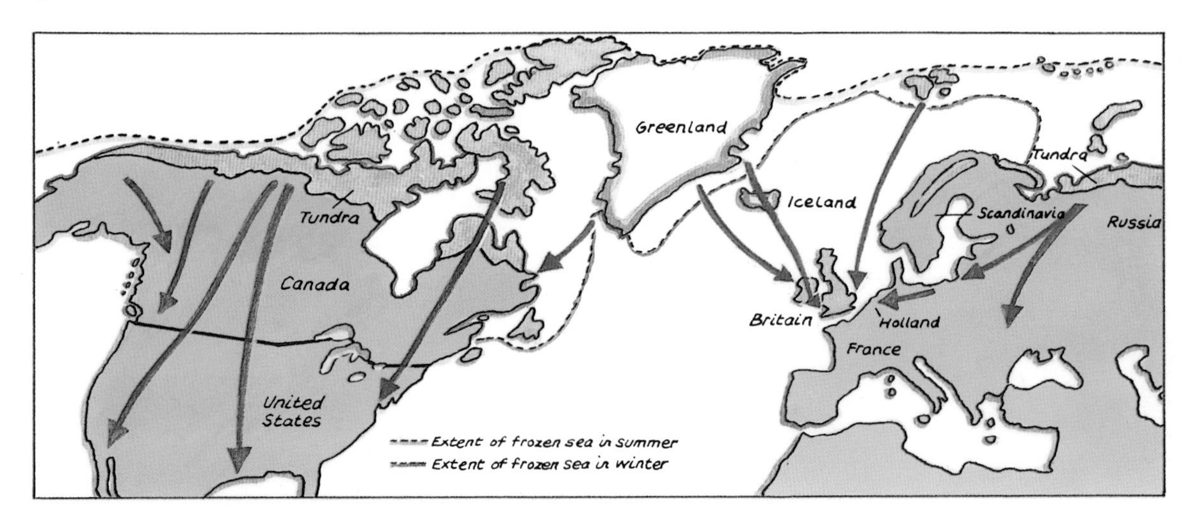

winters in Mexico, geese and swans also have different homes for the winter.

At the beginning of winter geese and swans migrate southward to coastal areas. The weather is always warmer by the sea than farther inland. Birds from the tundra of Canada migrate south to the coasts of the United States. Birds from Scandinavia and Russia migrate to Britain, the Netherlands, and France. Some geese and swans from Canada travel all the way to Britain. To do this they have to fly thousands of miles over the sea, crossing Greenland and Iceland on the way. The following spring, after spending winter in this southern home, they migrate all the way back to their tundra paradise.

Geese and swans migrate in family groups. The young birds learn to fly just in time to migrate with their parents. Usually, lots of families migrate together. A single flock

Swans fly with a powerful wingbeat, but they cannot maneuver well. Sadly, a migration can sometimes end by crashing into cables.

The pink-footed goose breeds only in eastern Greenland, Iceland, and Spitzbergen. In winter it migrates south in flocks of 1,000.

might contain hundreds of families.

Geese and swans migrate by night as well as by day. The sight of a migrating flock of geese or swans is spectacular. It flies in a special formation, in the shape of a V. One bird leads the way. The rest of the birds fan out behind this front bird to form the V-shape. The birds take turns leading the flock.

Geese and swans call to each other with strange "honking" noises while they are migrating. This honking sound is very loud. It can often be heard before the flock comes into view. Scientists still do not know exactly what the birds are saying to each other when they make this sound. Perhaps they are talking about how not to lose their way!

The Bronze Cuckoo

So far in this chapter we have looked only at birds that live in the northern countries of America, Europe, and Asia. But there are also birds that migrate between homes in the southern countries. One such bird is the bronze cuckoo.

In the summer months the bronze cuckoo lives in New Zealand. When winter comes, it leaves New Zealand and migrates northward over the Pacific Ocean. Scattered around this ocean are thousands of tiny islands.

The bronze cuckoo migrates from thousands of Pacific islands into New Zealand. Did this lead to one of the great human migrations? Some people think New Zealand was discovered by following the birds.

For the cuckoo, these islands make ideal winter homes. The weather is warm and there is plenty of food in the inland forests to last through the winter. In spring the cuckoos migrate back to New Zealand.

The bronze cuckoo is famous because it is thought that people discovered New Zealand by watching the birds migrate. About a thousand years ago, people living on the Pacific islands watched each spring as the cuckoos set off southward across the sea. Eventually they became curious. Where did these tiny birds disappear to each spring? Curiosity led these people to follow the cuckoos in their boats. After many days of sailing they reached New Zealand. These sailors were the first people to set foot on the island, and their Maori descendants still live there today.

5 Crossing the Equator

In northern countries each September, hundreds of birds can be seen gathering together on telephone and electricity cables. Perhaps you have noticed them yourself. Sometimes there are so many birds that the wires sag under their weight. Then, all of a sudden, they are gone. The birds you saw were probably swallows, and they were preparing for a very long migration that they take each fall.

All the birds in this chapter migrate very long distances. They are also special because they all cross something that we call the "Equator." What is the Equator, and why do some birds travel so far in order to cross it?

Why Cross the Equator?

Most of you know that the Earth is shaped like a gigantic sphere. The top half of the Earth's sphere is called the "northern hemisphere." The bottom half is called the "southern hemisphere." Where the hemispheres meet is called the Equator.

When it is winter in the northern hemisphere, it is summer in the southern hemisphere. January is the coldest month in North America and Europe but the hottest in Australia, New Zealand, and South Africa.

Most of the birds that cross the Equator feed on insects. There are far more insects in hot weather than in cold. It would be ideal for these insect-eating birds if they could live in hot weather all year round.

Swallows gather for their winter departure. They are packed so tightly on these telephone wires that there is hardly room for one more.

By crossing the Equator the birds manage to do this. In September they migrate across the Equator and into the southern hemisphere where the summer is just beginning.

The birds stay in the southern hemisphere until March. In March they migrate back across the Equator, just in time for summer in the northern hemisphere.

Birds that cross the Equator have

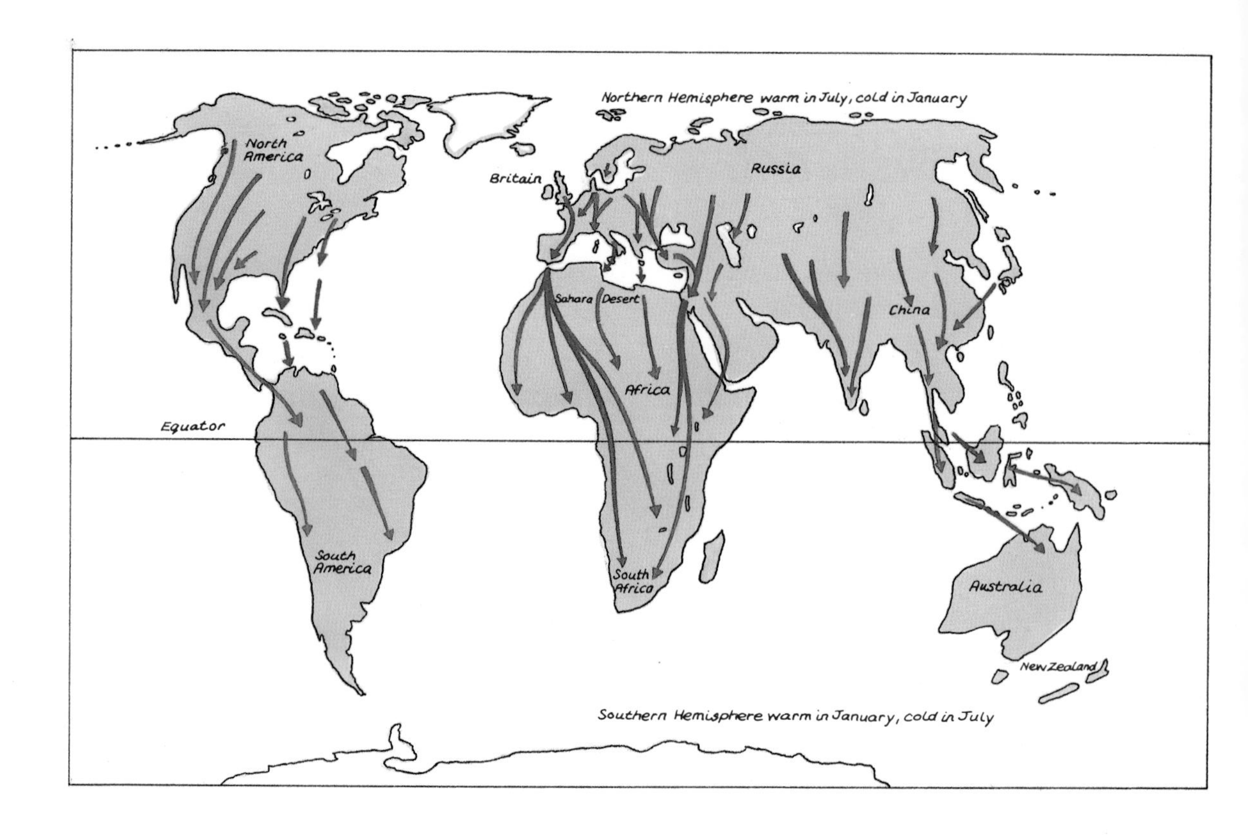

This map shows the migratory routes of swifts, warblers, and swallows. These birds are able to migrate 5,000 miles and back again.

two homes, one in each hemisphere. These two homes are frequently thousands of miles apart.

Warblers

Warblers are tiny, restless birds with dull brown feathers. They are champion migrants, though. Some, like the willow warbler, migrate as far as 5,000 miles every year. Each September this tiny bird migrates from its home in the northern hemisphere to its second home in the southern hemisphere. Then, the following March, the willow warbler migrates all the way back to where it started from again.

For a bird that weighs no more than a few ounces, this is an incredible distance to travel. To see how far the warbler's journey feels, you would have to travel to the moon and back again five times!

The Common Swift of Europe

Swifts are perhaps the most energetic of all birds. They spend much of their time flying around at great speed. Swifts are even able to eat on the wing. Most birds look for their insect food on the ground or up in trees. The swift, however, catches and eats insects on the wing. Any insect that flies too near this bird is likely to become its next meal! Like all the birds in this chapter, swifts migrate to ensure their supply of insect food.

Swifts have a home in each hemisphere. They spend from April to August in their northern hemisphere homes, and from October to February in their southern hemisphere homes. By migrating 5,000 miles or more from one home to another, these birds live in summertime all year long. This ensures that there are always insects for them to feed on.

These swift chicks will soon be living most of their lives in the air. They will feed, drink, bathe, mate, and even sleep on the wing.

Young Swifts

Swifts build their nests on cliffs or in the roofs of houses for their homes in the northern hemisphere. They lay eggs, and after about five weeks the chicks hatch. Swift chicks have a very special ability that is not shared by most other chicks. They are able to fly as soon as they leave the nest. Most young birds learn to fly by a gradual "trial and error" process. During this learning period some chicks have accidents and die. However, when the young swift is ready to leave the nest, it just falls off the cliff or roof into the air, spreads its wings, and flies. This is the beginning of the young bird's first migration.

A swift's first migration is a truly

amazing feat. The young bird flies continuously for two years without once settling for a rest. It does not even stop to sleep at night. Swifts do not settle on the ground or in trees to sleep like other birds. Instead, they climb even higher in the sky and spend the night alternately gliding and flying. Only when the bird is gliding does it catch a few moments of sleep. It has to keep waking up to make sure that it does not bump into any other sleeping swift.

Nests of the white-rumped swift of Africa and Asia. The Chinese encourage swifts to nest in their houses, believing they bring good luck.

In those two years between leaving the nest and the first time it settles, the swift travels over 18,000 miles by migrating twice to its home in the southern hemisphere and back again, all without stopping. Why does it stop after two years? Certainly not because it is tired. At two years of age the swift is old enough to build its own nest and lay eggs. To build a nest, the swift must fly into a roof and settle. When the eggs are laid, the bird must sit on the nest and keep them warm.

As the swift spends hardly any time perching, it has only tiny, short legs. Its wings, though, are magnificently long. The swift's legs are so short and its wings are so long that if it ever settles on the ground instead of a high perch it cannot take off again. If the bird tries to take off, its wings just flap uselessly on the ground. Sadly, if a swift lands on the ground it will eventually die.

The Barn Swallow

In the northern hemisphere, the barn swallow is found in North America, Europe, Russia, and China. Every year, throughout September and October, thousands of these birds leave their northern hemisphere homes. They set off on a journey that takes them across the Equator to winter homes in South America, South Africa, or southern Asia. For part of their journey, swallows migrate together in flocks of about

Swallows are clumsy on the ground, and rarely visit it. They prefer to perch up high, like this flock wintering in Kenya.

thirty birds. In these flocks young and old birds fly side by side. For some of the young birds it will be their very first migration. Let us imagine what the journey is like for a young barn swallow born in northern New York.

By September or October the young swallow is a good, strong flyer. He eats lots of food so that he has enough energy for his migration. As the time to migrate gets nearer, the young bird becomes restless. This is when swallows gather together on telephone wires or in groups of trees. Perhaps you have seen these flocks and wondered what they were doing. Then one morning they set off.

The young bird departs with the other swallows. The flock flies south across the United States until it reaches the Gulf coast. Here our young swallow will see the sea, perhaps for the first time. The Gulf of Mexico is a natural barrier and it is too wide for the swallows to fly across. Therefore they take one of two routes. Either they follow the island chains that ring the Caribbean Sea, or they fly along the land isthmus of Central America, which connects the continents of North and South America.

Like the swifts we looked at earlier, swallows are also able to feed on the wing. However, they do have to stop and land every night to sleep.

The young swallow will sample a variety of sleeping places on its migration. It may spend some nights on roofs or in the eaves of houses. Other nights will be spent on sheltered cliffs, in trees, or even perched on rushes and reeds in wetlands areas.

A Few Days' Rest

Sometimes the flock finds a place with lots of insects to eat, and will rest there for a few days. After some longer flights the birds are tired and have to gather their strength.

Most of the eastern coast swallows will follow the island route across the Caribbean, island hopping from one small land mass to the next from Florida down through the Greater and Lesser Antilles to the South American mainland in Venezuela. The western swallows proceed south through Mexico and Central America. In some places like Panama the isthmus is less than 50 miles wide, but the swallow is a clever bird. The older birds in the flock know that the strip of land is not wide. The younger swallows will simply follow the older members of the flock south into Colombia on the South American continent.

Southern Homes

Once the swallows arrive in South America they proceed south to their respective winter homes. Now, the birds that are going south have a much easier life, for they proceed through luscious rain forests of the Amazon River where insects are plentiful.

During the next few weeks our swallow crosses the imaginary line of the Equator. The end of the journey is almost in sight. Many birds winter in southern Brazil and northern Argentina. Some of the birds go as far south as Tierra del Fuego at the southern tip of South America. After eight or more weeks of migrating our swallow has finally reached its southern hemisphere home in November, perhaps 10,000 miles from where it started its journey.

The House Martin

In Europe the house martin is found in many of the same places as the swallow. These two birds can often be seen feeding together in one big flock. They also set off to migrate together. However, unlike swallow migration, the migration of the house martin puzzled scientists for years. After crossing over to Africa from Europe along with the swallows, this mysterious bird completely disappears!

For many years, nobody knew where the house martin went once it reached Africa. The whereabouts of its southern home was a complete mystery. Then one day recently there was a terrific storm over Zimbabwe in southern Africa. Huge hailstones fell out of the sky and crashed onto the ground. Down with them came a few unlucky house martins, knocked unconscious by the falling stones. The house martins' long-kept secret was solved. It seems that they spend the winter months high in the sky over southern Africa and, like the swift, apparently never settle anywhere to sleep.

Soaring Birds

Some birds, like the ostrich, are so big that they cannot fly. There are other birds too big and heavy to flap their wings for more than a few minutes. Storks, cranes, vultures, and some of the big eagles are like this. These birds use a very clever trick to help them fly long distances. They soar and glide like glider planes.

Wind does not only blow at ground level. Sometimes it blows straight upward. Birds like storks use this upward wind to take them high into the sky. Then they just glide around in circles, letting the wind blow them even higher. This behavior is called "soaring." When the birds are very high, they glide in the direction they want to go. They hardly ever need to flap their wings. They lose height only very slowly. When at last they have drifted too

The majestic golden eagle can soar for hours among the mountain currents. When it spots its prey it swoops with great speed and force.

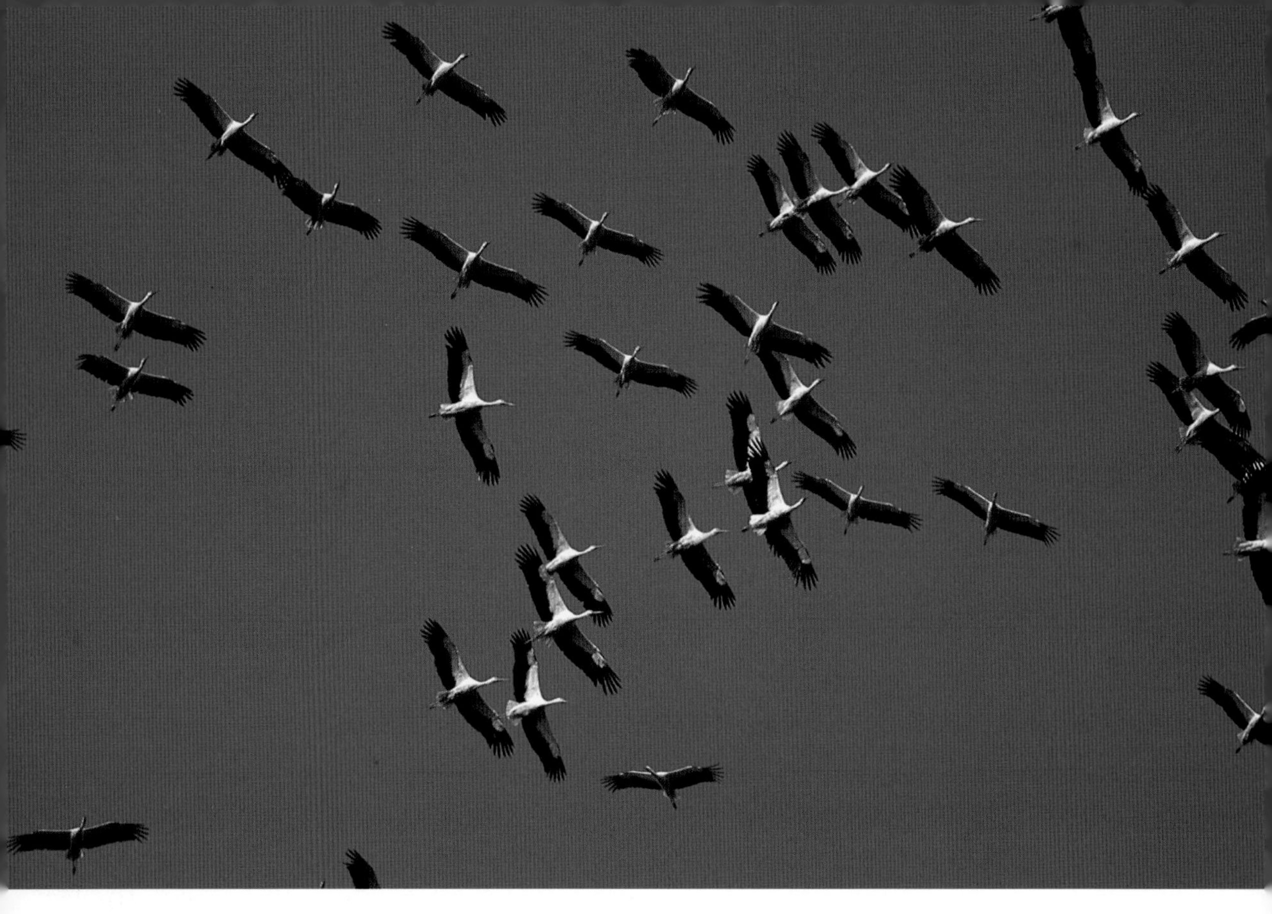

European storks make one of the longest migrations of the soaring birds. Large flocks depart from towns in Europe and fly the entire length of Africa.

low, they seek another upward wind. This is how European storks are able to migrate thousands of miles, letting the wind do most of the work.

Only if storks cross the sea do they fly into trouble. There are no upward winds over the ocean, so they have to use their wings like other birds. Sometimes they become exhausted, fall into the water, and then drown.

Storks breed mainly in the northern areas of Europe. They begin to migrate to their winter homes in South Africa at the end of summer. However, in their way lies the Mediterranean Sea.

There are two main routes to Africa. Storks of western Europe fly through Spain to Gibraltar, and cross the Mediterranean at its narrowest point. Storks of eastern Europe fly through Turkey, then south around the other end of the Mediterranean Sea into Egypt. Then they fly the length of Africa, a journey that takes six or seven weeks.

In February they leave their winter homes and arrive back in Europe eight weeks later. In parts of southern Germany, and in the Alsace region of France, the white stork is thought to bring good luck. Each spring people repair old rooftop nests, or even build new ones, to entice the storks to stop and breed there.

6 Seabirds

Nearly all the birds we have looked at so far live on the land. Some birds, however, prefer to spend their time on the sea.

The only seabird we have looked at so far in this book is the king of all wanderers, the wandering albatross. In this chapter we will look more closely at four more seabirds: the gannet, the Tasmanian muttonbird, the Arctic tern, and the emperor penguin.

The Gannet

The gannet is a large, elegant bird with black tips on its wings and a yellowish-colored head. The rest of its body is covered in brilliant white feathers. When a gannet hunts for food it flies along above the sea at a height of about 65 feet. Its eyes are fixed on the surface of the water as it watches for fish. As soon as the gannet spies a fish it goes into action. It folds its wings in a special way, then dives at great speed into the water. A gannet rarely misses a fish.

Gannets nest in big groups, called colonies, on rocky cliffs in the North

While cliffs are still being lashed by winter gales, male gannets return to take possession of the nests they left last year.

Atlantic. The parent gannets may fly as far as 60 miles from the nest in their search for foods for the chicks. They keep a special lookout for shoals of fish on these hunting trips. Shoals of fish are easy prey for these birds. They carry the fish back in their stomachs to the nest. When the young gannet feeds it actually puts its bill inside the parent's bill. The food is then brought up from the parent's stomach for the young gannet to eat.

Once the chick has grown and can fly, the parent gannets and their offspring migrate to a winter home. This winter home may be as far as 600 miles from the summer nest. However, gannets stay in the northern hemisphere. They do not cross the Equator in their search for a winter home. The following spring they return to exactly the same cliff.

The Tasmanian Muttonbird

The Tasmanian muttonbird builds its nest in burrows in the ground. Usually it digs its own burrow, but it will sometimes take over an abandoned rabbit burrow. This bird nests on the islands in the Bass Strait, between Tasmania and the Australian mainland.

The muttonbird begins to migrate in March, as soon as the nesting season is over. It migrates all around the Pacific Ocean. The journey lasts seven months and is 20,000 miles long.

First it flies toward New Zealand, then north toward Japan where it arrives in May or June. By the end of August it can be found feeding near the Arctic Circle in the Bering Sea. In September it sets off down the coast of Alaska and North

Another name for the Tasmanian muttonbird is the short-tailed shearwater. It breeds only on a small group of islands in the Bass Strait.

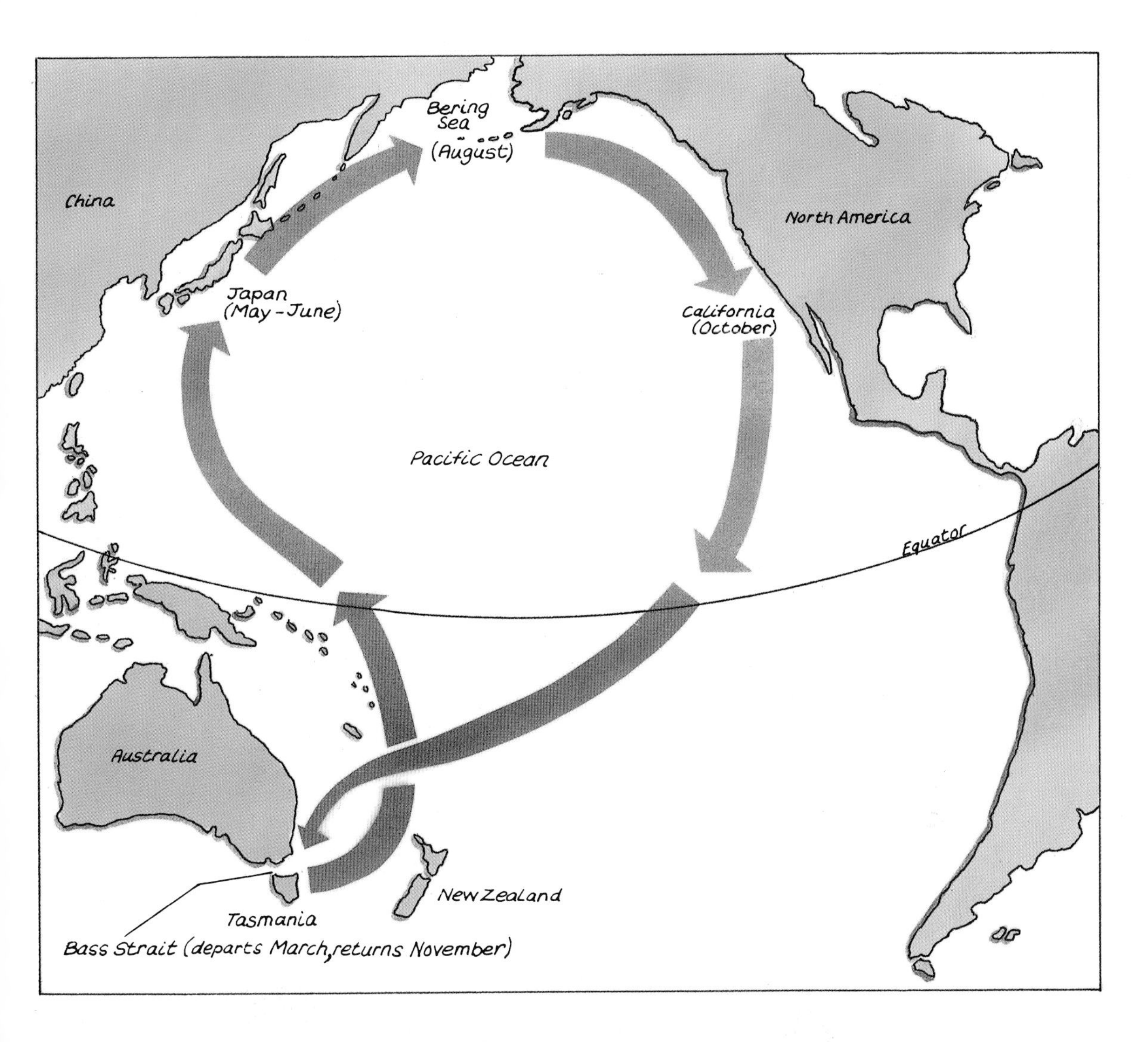

The extraordinary migration of the little muttonbird is shown on this map. Between March and November it flies all around the Pacific Ocean—a journey of 20,000 miles.

America. In October it leaves the coast of California and heads southwest, back toward Australia. Way out in the center of the Pacific it crosses the Equator.

The muttonbird's journey ends back at the Bass Strait islands in November. During the journey it feeds by flying close to the waves and snatching fish from the surface.

The Arctic Tern

This graceful bird looks like a small and delicate seagull. Sometimes the tern is called a "sea swallow," because it has a forked tail like the swallow. The Arctic tern is a very special bird. Every year it migrates farther than any other animal. It not only crosses the Equator but it very nearly flies all the way from the North Pole to the South Pole and then back again.

The Arctic tern breeds all around the Arctic Circle. In Greenland some

Arctic terns nest only a few hundred miles away from the North Pole itself. Their nests just lie on the icy ground. Arctic terns nest from April to August. For the rest of the year this tiny but incredibly strong bird is migrating. It feeds on the wing in the same way as the Tasmanian muttonbird.

The terns start their long migration as soon as their chicks can fly. This is usually during August. They start by flying south, down the coasts of North America and Europe. They cross the Equator in September or October, and continue down the coasts of Africa and South America. They keep migrating until they reach the great Southern Ocean where wandering albatrosses fly.

Here, in the Southern Ocean, the great albatross and the tiny tern feed side by side for a while. The terns may also meet penguins as they feed among the floating ice. Some of these tiny birds even reach Antarctica. Antarctica is at the opposite end of the world from the terns' nests that they left in the summer.

In February the terns begin the long journey back home to the Arctic Circle. They do not arrive until

The Arctic tern holds the record for the longest migration. It flies from one end of the world to the other, and back.

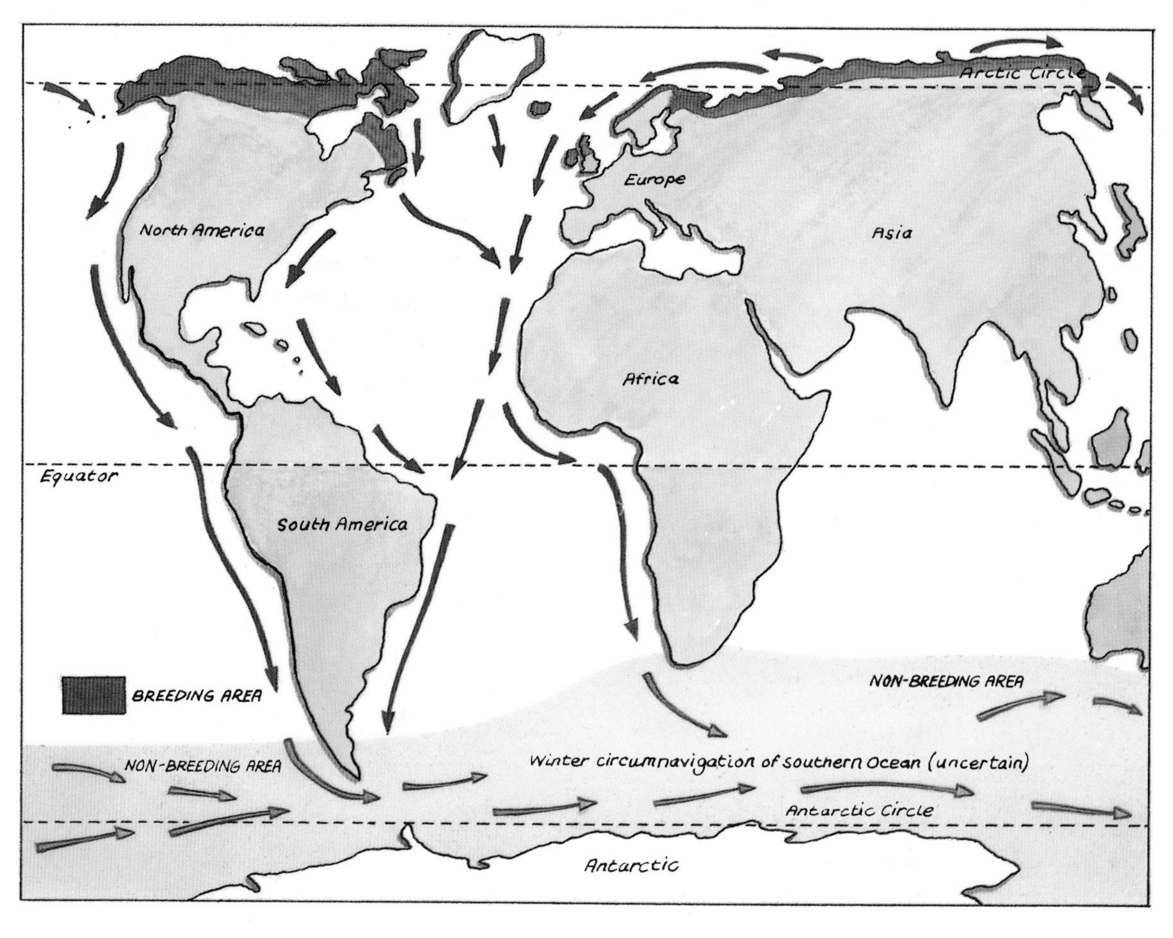

The penguin walks hundreds of miles to the sea to find food for its chick. The egg is not even hatched when it begins its journey.

April. Each year the Arctic tern spends about eight months migrating and travels about 25,000 miles. Among migrating birds, the tern holds the distance record.

Emperor Penguins

You are probably familiar with the awkward-looking penguin. Penguins live near the South Pole in Antarctica. They feed mainly on fish and tiny animals called "krill." Krill are found in the sea and look very similar to shrimp. Penguins are good at catching these tiny animals because they are excellent swimmers. Do not be fooled by the awkward way they shuffle along on land. Once they are in the water the penguins are as graceful as the otters.

The emperor penguin is the biggest of all penguins. It stands about three feet high. Penguins nest together in large colonies. Some colonies contain thousands and thousands of penguins.

The "nesting" season starts at the beginning of the Antarctic winter, but penguins do not actually build nests. Instead, the single egg is gently rolled onto one of the parent's feet, away from the frozen ground. The parent sits forward over the egg so that it is covered with warm feathers. Male and female parents

A five-week-old chick crouches on its parent's feet, off the icy ground. Soon it will migrate to the sea and spend several months there.

pass the egg between them and take turns sitting on it.

While one parent sits on the egg, the other has to walk to the sea to catch food for both of them. Groups of parent penguins form long columns and migrate to the sea together. To keep warm, each penguin presses closely against its neighbors. From a distance, this column of migrating birds looks like a troop of soldiers marching across the frozen wasteland. In winter, when the sea is frozen over, the penguins may have to walk hundreds of miles before they reach the sea.

After feeding themselves, the penguins set off on the long return journey. They carry food for their mates and chicks in their stomachs. If they get hungry they might digest a little of this food. The parent that stays behind might have to wait weeks, or even months, for its mate to return. When its mate arrives it regurgitates the food so that its family can eat.

In spring the ice starts to melt, and the penguins do not have to walk so far for their food. This is when the chick hatches. At first, only one parent at a time goes to the sea to catch food for the chick. The other parent stays behind to look after the young bird. When the chicks are bigger, they gather together with just one or two adults to look after them. Then both parents can go to collect the family's food.

The chicks grow very quickly. As soon as they are able to swim, they migrate to the sea with their parents. Like the great wandering albatross, the penguins wander around for a few months in the icy seas. Some emperor penguins travel over 300 miles from their breeding place during this wandering.

Although the emperor penguin is the biggest penguin, it does not migrate the farthest. The Gentoo penguin often migrates as far as 600 miles from the place where it nests.

7 How Birds Find Their Way

Most birds build a nest in which to lay their eggs and raise chicks. When any bird goes on a journey, it has to be able to find its way back to its nest. Some, like the house sparrow, have to return from only a very short distance. Others, like the Arctic tern, manage to find their way back after a journey of 25,000 miles! How do they do it?

A flock of colorful flamingos, which manage to look ungainly and graceful at the same time, take off from Lake Nakuru in Kenya.

Learning Landmarks

Young birds are very good at remembering places they have once visited. When they first learn to fly, they spend many hours exploring. At first, they do not go very far. As they grow older though, they start to explore farther and farther away from home.

As they fly over the landscape, the birds take special note of any hills, mountains, and rivers that they see. Things like these are called

"landmarks." Sometimes the young birds are taken on exploratory trips by their parents. The parent bird points out the most important landmarks to its youngster. The young bird learns to use these landmarks to find its way home. It remembers which landmarks are near its home and which are farther away. It also remembers the direction of the different landmarks from its home. After a while, the young bird is able to guide itself home by looking for each of the landmarks in turn.

Birds are also very good at remembering the way different places smell. These smells also help young birds to find their way home.

These young swifts will soon be ready to launch themselves from the nest and begin their first flight. It will last for two years.

Migrating for the First Time

Earlier in the book we looked at all sorts of birds that have two homes. The two homes are often a long way apart. The bird breeds in its summer home, then in autumn it migrates to its second home.

The first migration of a young bird is a remarkable feat. How does the bird know when it is time to migrate, or the direction in which to go? Also, how does the bird know when to stop flying? After all, a bird migrating for the first time has never seen the place for which it is heading. How does the first-time migrant even recognize its second home?

The "Internal Clock"

All migrant birds are born with a special sense. Scientists call it an "internal clock." This "clock" does not only tell the bird the time of day or the time of year. It makes the bird do things it needs to do without thinking about them. For example, when the weather gets colder, it makes the bird want to eat lots of extra food. The bird needs this extra food to give it strength for the journey. A bird's internal clock also tells it when to set off, which direction to fly in, and when to stop flying. All birds, young and old alike, use their internal clocks when they migrate.

Learning from Older Birds

Besides using their internal clocks, young birds learn how and when to migrate by watching older birds. Some young birds are able to watch their parents. Others, which fly alone, just have to keep a close eye on their neighbors.

Just before migrating time, all birds are "told" by their internal clock to eat lots of extra food. For young birds, seeing their neighbors eating lots of food is an extra reminder for them to do the same. When the time comes to migrate, the young birds just follow their elders. Most birds migrate in large flocks. Some though, like geese and swans, often migrate in smaller family groups.

On their first migration, young birds are careful to notice and remember all the big landmarks they see. They are also very inquisitive. If they fly over something that looks interesting, like a good place to eat or sleep, they will often fly down to investigate.

Sometimes they even fly back a short way to have a second look at interesting places. When this

A pair of common cranes, each of which looks like a reflection of the other. During their migratory flights they travel at high altitudes—possibly up to two miles.

happens, the youngster is often left behind by the main flock. This is not as disastrous as you might think. The youngster just sits tight and waits for another flock to come along. It then joins this new flock. Young birds on their first migration often drop in and out of different flocks. Those with their parents, however, are not allowed to be so curious. They only stop when their parents do.

Every so often, the flock or the family group stops for a few days to eat lots of food. The youngsters remember these good feeding places. Finally, the young bird arrives in its second home. How does it know when it has arrived? First, the young bird's internal clock tells it that it is time to stop migrating. Also, there will probably be a lot of older birds around that have already arrived before this bird.

A pair of young terns learning to fly. Terns are impressive migrants. In a few weeks these birds will be thousands of miles away.

Heading back is easier. Again, the young bird's internal clock tells it when it is time to get ready to fly back. Also, it will notice some of the older birds beginning to migrate. This time, though, the bird is familiar with much of the land it flies over. It recognizes the landmarks that it saw on its way out. Even so, it still travels with the rest of the flock. Eventually, the youngster recognizes the place it explored around the nest in which it was born. The young bird realizes that it has returned to its original home.

Direction-finding

Birds learn to recognize the different directions when they are very young. There are plenty of clues to help them tell one direction from the other. Most birds are capable of using all these clues. First, there is magnetism.

Our planet is like a huge magnet floating silently through space. It is surrounded by strange forces called "magnetic lines of force." These are the forces that make the needle of a compass always point north. All birds can sense magnetic lines of force, and they use them to tell

Swallows do not always travel together. Some can arrive weeks early. Hence the expression "One swallow does not make a summer."

which way is north. So they have a sort of compass inside their heads.

Birds also use the sun, moon, and stars as direction signs. Young birds are very alert and observant. They notice that the sun rises in the east every morning, and sets in the west. Birds in the northern hemisphere notice that in the middle of the day the sun is always exactly in the south; and in the southern hemisphere it is exactly in the north. They know that it is south by using their ability to sense magnetism; and they know it is midday because of their internal clocks.

All of these things are learned during the first few weeks after young birds leave the nest. This is the time when they are exploring. By the time they are ready to migrate, they can tell which direction to go by using the sun, moon, stars, and magnetism. So they know which way to fly whether it is day or night, sunny or cloudy.

It has taken scientists a long time to find out how birds avoid getting lost on their long journeys. But in spite of our knowledge, we are still full of admiration. These delightful creatures can do so many things that we cannot. Flying is perhaps the least of their wonderful skills.

Glossary

Antarctic The whole area of the globe lying to the south of 66.5 degrees south latitude.

Arctic/Arctic Circle The Arctic is the whole area of the globe lying to the north of 66.5 degrees north latitude. The Arctic Circle is the line drawn on a map or globe which marks this boundary.

breeding Producing and rearing offspring.

climate The general weather conditions usual for the time of year and for the region. What a bird eats depends on the climate.

egg The hard, oval-shaped shell from which young birds are born. The egg is produced from the mother's body. When the bird is ready to be born it pecks its way out of the shell.

Equator The (imaginary) line around the center of the globe separating the northern and southern hemispheres. Most of the Earth's hottest regions are at sea level on the Equator. There is no winter or summer.

first home See **home.**

grassland Land on which grass is the main vegetation, like the American prairie or the East African plains.

home The habitat where a bird makes a nest, lays eggs, and raises its young is its *first* home. The place it migrates to is its *second* home.

homing The ability of a bird to find its way back home after traveling great distances.

internal clock The instinct that tells a bird when it is time to migrate.

migrant/migrate/migration The habit of moving from one habitat to another (usually in search of food) is called *migration*. An animal that *migrates* is a *migrant.*

nest A bed of moss, twigs, leaves, etc., in which birds lay eggs and raise their young. It is not a "home" because it is not used except in the breeding season. Not all birds make nests.

north The direction a compass needle points. Most maps are drawn so that the northernmost part is at the top.

northern hemisphere The part of the Earth north of the Equator.

North Pole The northernmost point of the Earth. It is surrounded by ocean covered with permanent ice.

offspring The young birds produced by one pair of parent birds.

regurgitate To vomit up partly-digested food from the stomach so that it can be eaten again (usually by offspring too young to find their own food).

second home See **home.**

soaring Drifting and rising in warm air currents. Once high in the air the bird can begin to glide.

south The direction opposite to north.

southern hemisphere The part of the Earth south of the Equator.

South Pole The southernmost point of the Earth. It is surrounded by land covered by permanent ice on which nothing grows.

tundra A vast area lying south of the Arctic where no trees will grow on the frozen subsoil. The word also describes the scant type of vegetation that grows there in summer, and above the tree line on mountains.

wanderers Birds that spend most of the year migrating within one area, instead of from one area to another with the seasons.

wingspan The distance between the tips of a bird's two wings when they are fully spread.

winter The coldest season of the year, with little food available for birds: December-February in the northern hemisphere, June-August in the southern hemisphere.

More Books to Read

MIGRATION
Adult books that older children would enjoy.

Bull, John & Farrand, John, Jr. *The Audubon Society Field Guide to North American Birds: Eastern Region* Knopf 1977

Udvardy, Miklos *The Audubon Society Field Guide to North American Birds: Western Region* Knopf 1977

NATURAL HISTORY
Children's books containing some information on migration

Austin, Oliver L., Jr. *Families of Birds* Western Publishing 1985

Bailey, Jill & Seldon, Tony *Birds of Prey* (Nature Watch Series) Facts on File 1988

Bender, Lionel *Birds and Mammals* (Today's World Series) Franklin Watts 1988

Boy Scouts of America Bird Study Boy Scouts of America 1984

Burnie, David *Bird* (Eyewitness Books) Knopf 1988

De Jonge, Joanne E. *Of Skies and Seas* (My Father's World Series) CRC Publications 1985

Farrand, John *Eastern Birds* McGraw-Hill 1988

Hendrich, Paula *Saving America's Birds* Lothrop, Lee & Shepard 1982

Hirschi, Ron *The Mountain Bluebird* Dodd, Mead 1988

Johnson, Sylvia A. *Albatrosses of Midway Island* Carolrhoda Books 1989

Losito, Linda, et al. *Birds: The Plant- and Seed-Eaters* (Encyclopedia of the Animal World Series) Facts on File 1989

McPherson, Mary *Birdwatch: A Young Person's Guide to Birding* Sterling Publishing 1989

Parnell, Peter *The Daywatchers* Macmillan 1984

Selsam, Millicent E. & Hunt, Joyce *A First Look at Owls, Eagles, and Other Hunters of the Sky* Walker 1986

Wallace, Ian, et al. *Bird Life* (Mysteries & Marvels Series) EDC Publishing 1985

Wiessinger, John *Birds—Right Before Your Eyes* (Right Before Your Eyes Series) Enslow Publishers 1988

Picture Sources

ARDEA: Uno Bergrens 29; G. K. Brown 34; Graeme Chapman 26 upper; John Daniels 11; Eric Dragesco 33; Trevor Marshall 15; E. Mickleburgh 16; Bernard Stonehouse 39.

Bruce Coleman: Francisco Erize 19; C.B. Frith 30; Udo Hirsch 44; M.P. Kahl 41; Dieter and Mary Plage 20; Hans Reinhard 18 upper; F. Sauer/J. Kankel 43; Kim Taylor 7; Roger Wilmshurst 42; Joe Van Wormer 27 and 45; Konrad Wothe 31; WWF/Eugen Schuhmacher 40.

RSPB: S. and B.A. Craig 25; C.H. Gomershall 8, 13, 17; George McCarthy 22; P.R. Perfect 24; S.C. Porter 9, 10; M.W. Richards 12; E. Wright 35.

Simon Girling Associates/Richard Hull: 6–7, 14, 18 lower, 21, 23, 26 lower, 28, 37, 38.

The Wildlife Collection: Gerry Ellis cover.

Index